Questioning Extraterrestrials

And

Earth

## Acknowledgements

This booklet is dedicated to anyone who finds extraterrestrial findings and information on earth interesting. Thanks for reading.

Sources:

- usatoday.com
  - Universe is 13.8 billion years old, scientists confirm
- nationalgeographic.org
  - Age Of The Earth
- Google.com/images
  - Area 51 Sign
  - Alien Artifacts
- Wikipedia.org
  - Roswell Incident

Published on - 5/13/21

Our universe is about 13.8 billion years old, and the earth is estimated to be about 4.54 billion years old. Beings from other worlds have been a superstition for much of human history. From cave paintings to controversial videos, other beings have been questionable for thousands of years.

Even in modern times arguments for or against the existence of other beings still stand.  Without official government confirmation of the existence of other beings, everything speculated is up for debate.  Even though both sides of the argument do have impactful stances, the truth is still unknown.

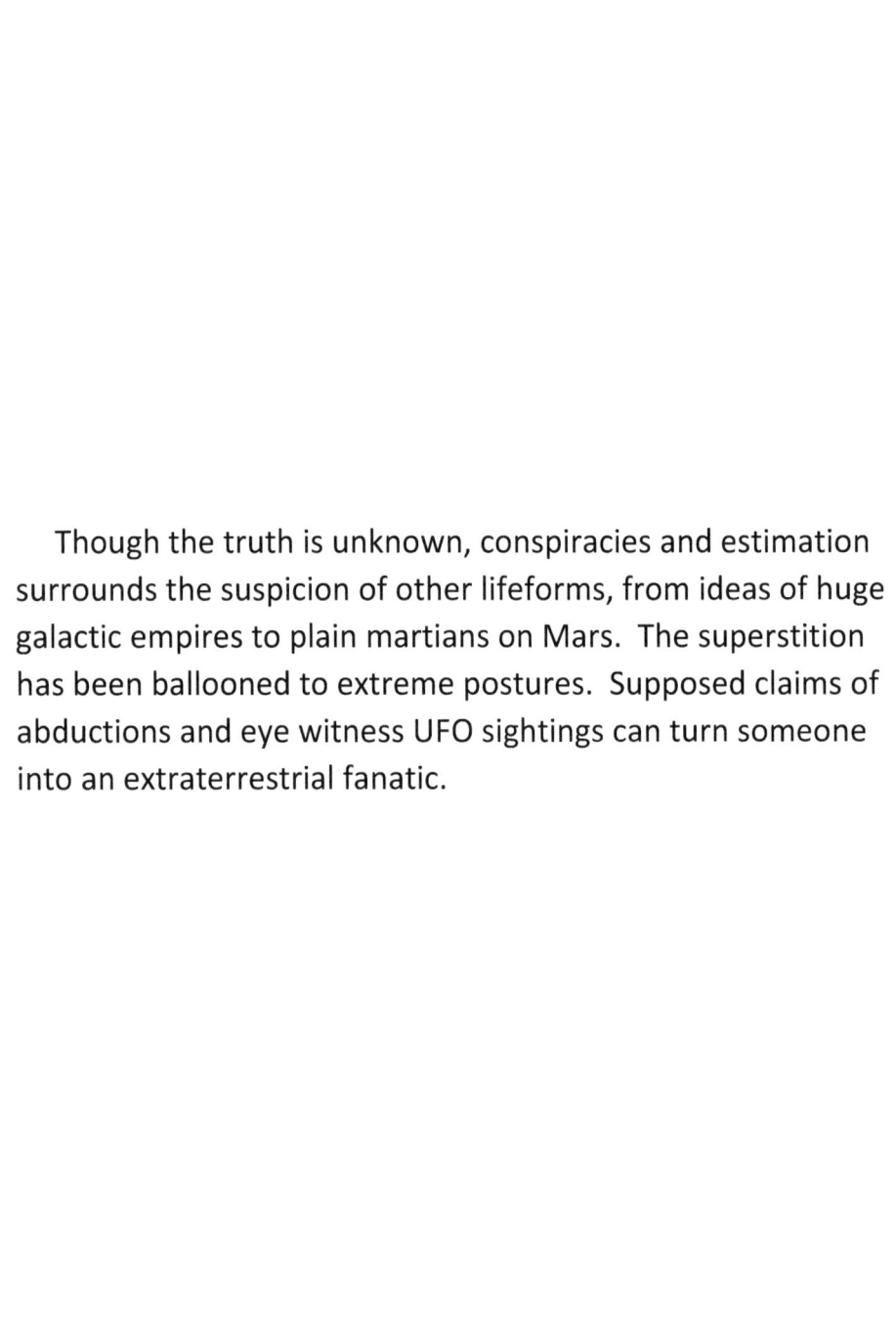

Though the truth is unknown, conspiracies and estimation surrounds the suspicion of other lifeforms, from ideas of huge galactic empires to plain martians on Mars.  The superstition has been ballooned to extreme postures.  Supposed claims of abductions and eye witness UFO sightings can turn someone into an extraterrestrial fanatic.

However, hunting around the world for other lifeforms is not the smartest or easiest thing to do and getting abducted on purpose is nearly impossible. Finding extraterrestrial life on earth is like looking for a light switch in a dark room. You could call it luck or an "out of this world" chance encounter.

High level government facilities like Area 51 are presumed to be looking into extraterrestrial life and technology. Of course, no one really knows what goes on in the airbase or these facilities, because access is restricted to civilians and even most government officials.

# WARNING

## MILITARY INSTALLATION

OFF LIMITS TO
UNAUTHORIZED PERSONNEL

AUTHORITY: Internal Security Act
U.S.C. 797
PUNISHMENT: Up to one year imprisonment
and $5,000 fine.

Ancient scribings, paintings and artifacts from past civilizations have been found all around the world and are being studied by officials.  They are decoding their meanings and origins in pursuit of finding extraterrestrial signs.

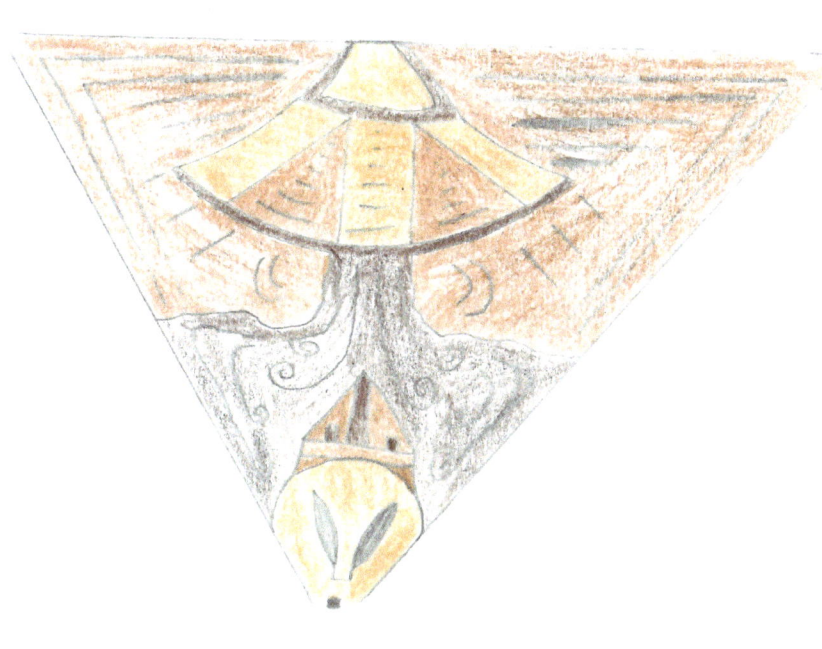

There have been rare cases of suspected UFO crashes in certain places around the world.  Claims like these are usually investigated by officials and are presumed to be covered up with a false story.  An example of this would be the famous "Roswell Incident".

Though no one can truly confirm the existence of extraterrestrial beings, all the attention and mystery cannot be denied.  The matter will be an ongoing exploration for years to come, or at least until the authorities say otherwise. Everything else surveyed comes down to personal opinion.

Questioning extraterrestrials and earth is a booklet speculating the presence of other beings on earth.

www.ingramcontent.com/pod-product-compliance
Lightning Source LLC
Chambersburg PA
CBHW041949240526
45473CB00036B/2862